THE LAST AMERICAN POEM

FLIGHT

by Charles Tidler

Drawings by
Johnny Tens

PULP PRESS

Titles by Charles Tidler:
North of Indianapolis
Straw Things
Whetstone Almanac
FLIGHT: The Last American Poem

THIS IS A PULP BOOK
Published by Pulp Press
Box 48806 Stn Bental
Vancouver Canada

ISBN 0-88978-024-2

A Canada Council Grant in the summer of 1975 kept the wolf from the door during the composition of "Highway 80 West." Portions of "Highway 80 West" originally appeared in *The Wormwood Review*. "The Pivot", in a very different version, was published in *Empty Belly*, and the poem "Flight" appeared in *3¢ Pulp*. — C.T.

Cover Photo by Jim Frew

Printed and bound in Canada.

This book is for

Billy

&

Brenda Jo

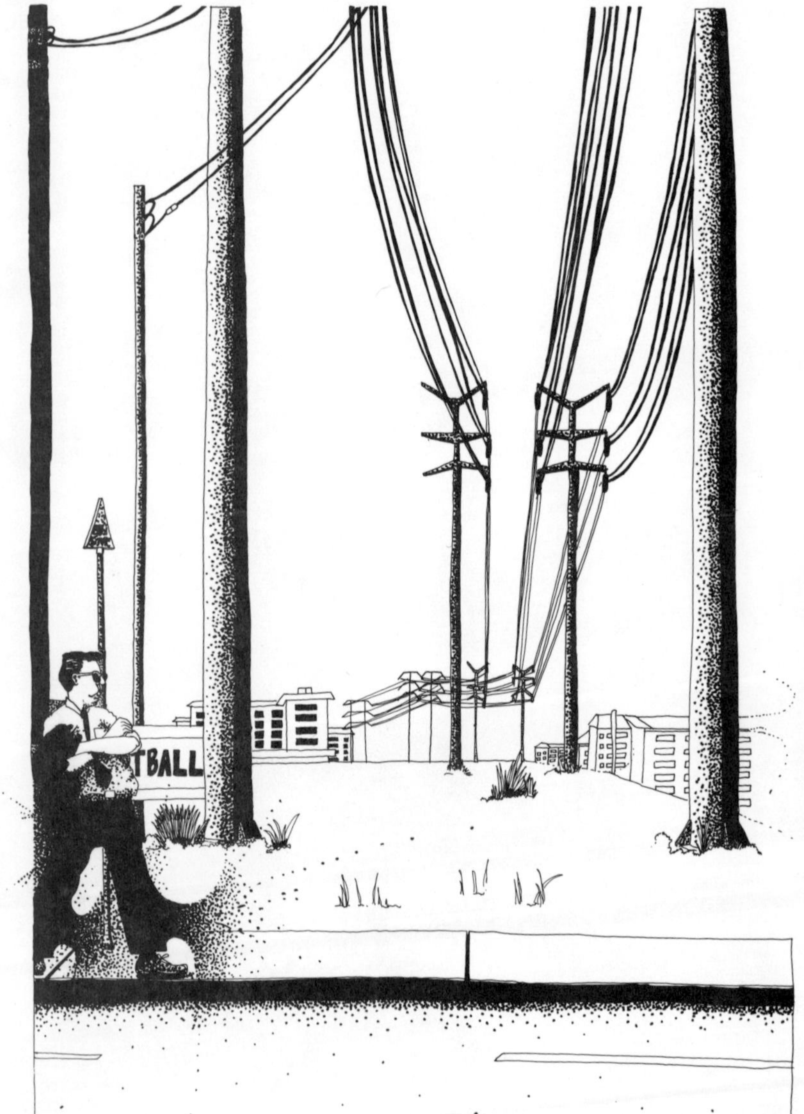
TBALL
TENS /76

Table of Contents

The Pivot

walked in a parking lot
a ball of newspaper
burning in a trashcan
yellow fire

about the petals of a black rose
beside a stump
six feet high
four feet across

sat upon it singing songs
and love poems
echoing off a wall of sound
New Babylon

typewriters raping
radios looting
phones blowing up whole cities
T.V.s belching and farting

IF YOU ARE A NATIVE-BORN UNITED STATES CITIZEN
CHECK BOTH APPROPRIATE BOXES
AND LEAVE REMAINING BLOCKS BLANK
IF YOU ARE A NATURALIZED CITIZEN
LIST CERTIFICATE NUMBER IF AN ALIEN LIST
YOUR ALIEN REGISTRATION NUMBER
IF YOU ARE STATELESS SO INDICATE YOUR
LOCAL BOARD WISHES YOU EVERY SUC—

I toss all the wishes
to the burning rose,
walk the road outside of town.

AN ALLEGORY

When all the world was drinking blood
From the skulls of men and bulls
And all the world had swords and clubs of stone

—Vachel Lindsay

AN ALLEGORY

When all the world was drinking blood
From the skulls of men and bulls
And all the world had swords and clubs of stone

—Vachel Lindsay

Highway 80 West

        The highway unrolling
    a ball of sun
slipped from the lap of a Muse
knitting a sweater of the heavens,
unrolling down the length of her green thigh
the bob of the heads of the wayside grass
hearing the highway swallow
sweet as tin whistle
follow, follow

the road thru *Steel Hollow*, thru *Black Grass*,
Pork Back, *Deadview*, *Lost Hope*

hearing the highway swallow
rise up!
follow, follow

thru *Backstab*, *Crass City*, *Oblivion*

and came to the river
following.

•

I wander along the river bank,
eyes withered by billboards
flashing girly-cues and cancers.
I would siphon a poetry
from the pool of our history
but only choke
on a mouthful of gasoline
to measure the octane
of our burned-out versification.

Take the whole
continent

and ask
no favors;

I wish never
to see peace

til we do:
Henry Clay.

The hi-rise apartments across the river rise,
bridges leap steel springs of reality,
wheels spinning thru the towns
strung like barbs
along the barbed wire constellations
of the interstate,

the river arthritic
and the land gone as smoke.

AMERICA LOVE IT OR LEAVE IT

•

The silver strings of the highway
stretch taut as lyre, a blur
of pastourelle
at 70 miles an hour.

Tractors plow the earth,
white clouds furrow the blue sky.
Orchards, ablaze,
pirouette upon the green hills
as the interstate tangles to a clover leaf,
antiseptic parking lot
and shitters.

•

The sun glows, a live coal
in the pit of a passed pipe
laid aside and growing cold
on the dashboard of the horizon.

The arroyo of a parched canto
on the cracked
lips of a gas jock,
"Fill 'er up?"
                    Local yokels
cruising up and down, back and
forth, north, south, east
and west the wide sundown streets
of Ogalala, Nebraska.
                              Here, Black Elk
once stood
at the center of the
world
        and wept for the people.

the peace broken
        like a stick

•

Ma dead,
the barn burned,

the war done,
me, two brothers and Pa

dipped our sorry butts
in alcohol,

crossed the Mississippi
in an old tin pan

riding Hell-for-leather
Cheyenne...

Bracelets of machinery,
monuments of bone
and all the fine dandy highways
running like shuttles
thru the tattered weave of the continent
let to rot on the broken loom.
I take it up,
stinking, by the handsful
and drive away.

•

The slide thru Salt Lake City, dusk,
coast a hundred miles
silent, lost to
contemplation, the husk of meaning
shed of clothing upon the prow
of a ghost ship,
ghost sea. Leviathan swims
upon the gold crests of the headlights
surfing a landscaped holocaust.

Sun glances off a mountain's shoulder,
salt flats
white as powdered bone
sifted to the bottom of a blue bowl
where a ghost phonograph
sticks like a pig—

"and ask no favors, and ask no favors,
and ask no favors, and ask..."

Heel your sign, Gemini, Christian, Fascist,
giggle into the void
suddenly made ridiculous as the first
astronaut to piss on the moon
or should the President miraculously appear on color T.V.
attempting to shake hands with the
Kosmos.

We can disappear.

•

Swig a porcelain cup
    of hot black coffee,
Eat a slice of pie
    the size of a splayed hand,
Light the tenth
    Lucky of the dawn

And jam a quarter
    into the jukebox:

        (((my name
                        yeah yeah yeah
is a household
                    money)))

The people in the *Neon Cafe*
    orange neon skin
    pink neon hair
are used to stretching fence,
rounding strays
or drifting in a high cab
running cargo K.C. to Frisco...

while outside,
in the drunken dazzle Reno noon,
a black leather gang
sporting the colors of *Hell's Politicos*
rough up Vachel Lindsay
for peddling his *Poems for Bread*
in the door-
way of the *Neon Cafe.*

Ritchie D. indents Lindsay's forehead
with a pair of brass knucks,
"There's your *Yes, The People*, punk!"

Vachel protests, "Sandburg said that;
I'm for *'Bryan, Bryan, Bryan, Bry—'* "

"Don't bullshit us," yells Lyndie J.
as he takes a mean bite at Lindsay's foot,
spits out pieces of boot,
toe and blood. "Argh,
all poets taste the skunk!"

Dicky N. rolls with laffter
on his sides upon the sidewalk,
pops open a brown bubbly
using Lindsay's mouth for an
opener, " *'All poets taste the skunk'*,
yuck yuck, you tear me up."

•

A white city burning
or a red city
burning or a black
city burning or a
yellow city burning or

crocus, anemone, dog-tooth-violet
flare amid the greenery,
the granite rock
bouquet of the High Sierras

where I look out
as the ribbon of highway descends...

or a bare 40 watt bulb burns
in the basement room
of a house hunched-up beneath an overpass
where a man cranks out the leaves
of a mimeo magazine.
He waves his bloody liver in the air,
a red book of the verbally mad
damned with the cultivation of the language,
and rails against the creeps, the punks,
the cheaps of a literature floating toward a mountain
of paper shredder fodder.

He quaffs the long beers
cold as the icy thigh
of an indifferent Muse,
flings molotov haiku

folding folding folding

the pages falling
one upon another page
as waves spilling their white hands
upon a beach
where the highway ends
and crumbles like cake in a child's hand,
cake in a child's hand.

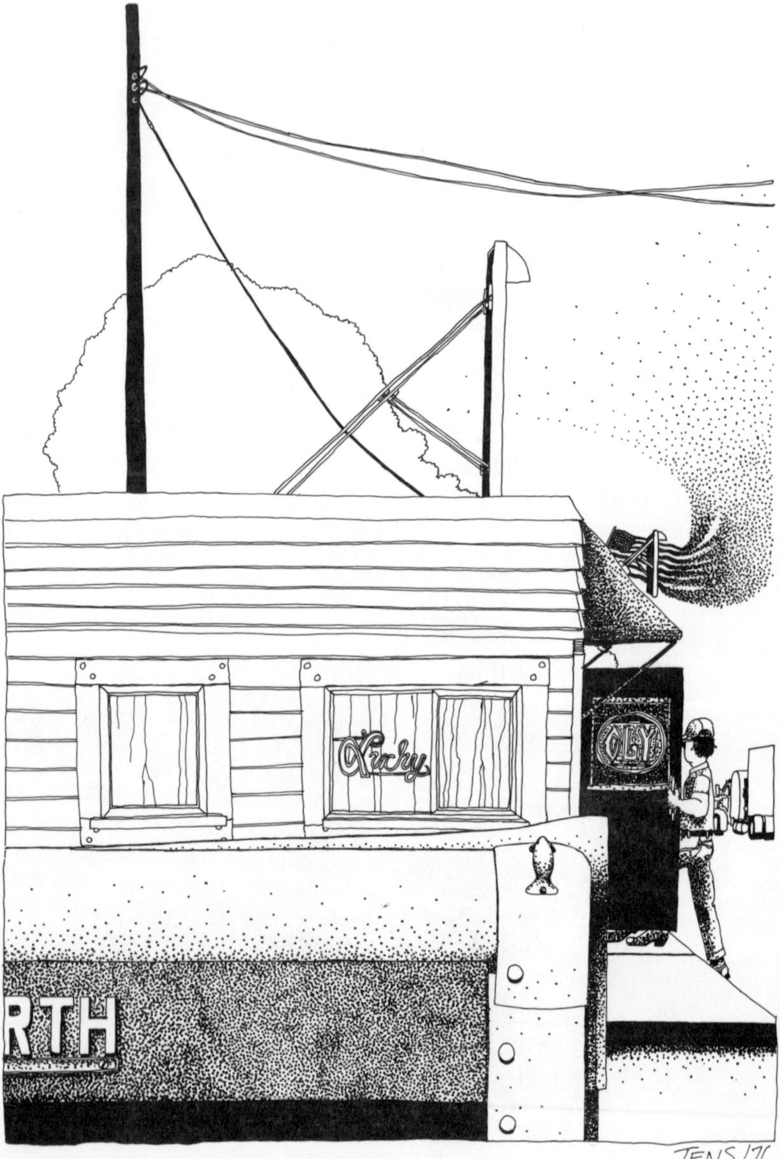
Lucky
OLY
RTH
TENS/76

Spilling The Beans Vancouver

"Okay, buddy,
the end of the line;
this baby's goin nowhere."
I collect my bags heavy as two bombs
from the belly of the coach:
ghost clothing,
four dozen spineless books,
shrapnel of song...

The hot air's a plastic sack,
sucks my face—
cigarette tastes like chalk,
the pssst of a lit fuse.
I slide
on a thin film of gasoline
towards the *Men's Room*.
My bags explode.

Whitman, Donne and Baudelaire
dance thru the lobby
arm in arm
with my tattered underwear.

Thoreau trips a newsboy.
Eliot and Blake smash a man's plate
in the coffee shop,
"Hey, I ordered hamburger steak..."

I follow Poe into the pisser.
Villon bangs on the door of a pay toilet.
A crippled old man
corners Rimbaud in a urinal.
Bukowski shits in a sink.
Burroughs shoots up in the wastecan.
Catullus scribbles on the wall:
SIC TRANSIT GLORIA MUNDI.

Outside in the waiting room
my bags have been cordoned off:
holey socks arrested,
T-shirts held for questioning,
my notebooks in a straight jacket.
Police wrestle Pound for a megaphone,
"WHAT THOU LOVEST WELL..."
Stevens rides upon a ripe peach
big as an elephant,
overturns a 40 foot bench.

Jeffers takes the next bus out of town,
Rexroth throws a haymaker at anybody;
Williams cackles with glee
as Dickinson strips

accompanied by Lindsay's sousaphone,
Sandburg's tambourine.
Would you believe Melville on vibes?

I do my famous chicken
shagass Hell-to-pay hundred yard dash
(without sock, shirt or verse)
jump a cross-town bus. Pound,
going down hard, echoes
in my ear:
the rest is dross

Wreck Beach

tea
marmalade
red ant

Red-headed sea witch rocked me
all night in the driftwood hut.
She's gone at dawn
with my last dry bit of fresh tobacco.
Strands of spring water
shape the sand
flowing of a woman's hair.
I dip cupped hands
and drink deep.

shadow
meat
beach

Tearing away from the binding
like a butterfly's wings,
I roll cigarettes from the mouldy pages
of a paperback *Gita*,

gather chips
of driftwood in a discard pail,
eat a hard corn cake
burnt black on red coals.

headstand
old text
footprint

A wind blows I have to lean to one side
to walk anywhere:
the turn, glide and vibration
of the breakers.
"I am your green orphan,"
rolling upon me
taking the sand between my toes.

moon
pebble
wine bottle

The mouse found pilfering
my breadcrusts this morning
chewed a hole thru my billfold
to lunch upon my name.
What can I say?
We're the best of friends.

ocean
basket
dance step

Rain washes me out,
high tide takes my shack down.
Eagle steps across the sky,
says to me,
"You do talk a lot of shit, boy."

nicotine
caffeine
downpour

Ironhorse Drifting

"Dad put me to th plow
        by seven, furrows of blisters
to my hand. What do th youth
        want today?
I was pulled outa school.

"H/A/R/D/W/A/R/E, th man sd,
        spell it B/I/G,
learn th A/B/Cs of calico,
        shoe leather,
barbed wire. Winnipeg, 1934.

"The Big War, quartermaster Africa,
        saw Mussolini swing,
th first boat home: S/O/C/I/A/L
        C/R/E/D/I/T, th man sd,
God'll look after th government,

"tend yr store. I worked hard,
        fingers to th bone,
voted straight, praised th Lord.
        What's wrong today?
What happened to a day's work

"for a day's pay? You wanta nother
    drink? I can't sleep,
th wife snores like a bloody train.
    Hey, get it?
Th wife snores.... Ha. Ha. Eh?

"I could afford to fly,
    but two days give a man
time to drink. Drink? I mean think—
    must be gettin tired.
Dad hada good team of Clydes

"in 1928. Know what a man made
    after a harvest?
Wanta know what I got? Whata you
    kids want, for Chrissakes?
Damned if I know; I know what's..."

Honor and Decay are brother and sister,
    hatch from one egg,
spawn Treason.

Each is ever
    the other's shadow.

•

The mythic snake of fertility
brought to fruction
by a thousand trampled tongues
(10,000 years)
is no match for the C.P.R. deisel
parting the thighs of a mountain
outside Golden, B.C.,
and climbing erect into the sky
as the big dipper swings as a gate hinge
upon the pole of the heavens:
I nurse a nightmare hangover
in the geodesic domecar,
a black hole
turned inside out
as a pocket of bone soot,
star ash—
my head in flight
purple flame flared to a fuse.

Honor traps me in the washroom
        with a razor. Decay throws her shoulder
to the door, the latch snaps shut,
        the cold tap runs
                into an open drain—
dawn thru the milky window,
        a slit thru a jugular.

•

"See the tracks a mile below
we wound over an hour ago?"

Night stars fade
and fall, snowflakes fly:
Ironhorse drifting
in the realm of dream and cloud.
Tell me a story
of white geese over blue lakes,
mountains higher than the Fists of Power,
sheer cliff face cataracts,
motionless, falling.

The engineers constructed bridges
where no bridges could go.

Little old lady
sits next to me in the domecar
crossing The Great Divide,
tells me of a Tower of Man
and a Poet's Bell:
the hearing is a ring of Beauty.
She must be crazy, cracked
like a dull bell, shows me her verses
dedicated to Daphnis and Thyrsis.
I hold up my hands, positive
and negative poles of an energy;
she winds her wool from me

into a ball, knits,
snips...

They drove tunnels through mountains
where no tunnels could drive.

Honor in a tux, pink cummerbund,
and a bright carnation of flesh
tricked me.
I took him for the maître d',
belched politely
turning over my water glass, nodded yes
thru a ketchup-stained napkin
at his invitation to meet the chef,
a three-legged sow in a chef's white mushroom hat
blowing her swollen nose into the soup pot.
Decay turns on me,
her blunt boar's teeth
a twin bull's-eye on my balls.

•

The Glorious City. Pastoral Country.
The wild reaches of the imagination,
the real into which we enter and revere
is necessary upon self-help, mutual aid
self sacrifice, mutual store

the hand on the hoe
and the hand on the sword
hand of water
hand of stone

twelve policemen dead in the snow

10 days of fasting and prayer
to invite the Will of God:
Riel announces He will perform
the Mass Himself.
"Rome," He cries, "has fallen!"
while Poundmaker, the Cree, loots houses,
dances in white women's stolen gowns
and burns a town.

Alone, Gabriel DuMont turns the reins
rides out of history
one snapshot carbine across chest
x'd with bullets over thick, glossy hide
rides out of civilization
to follow buffalo
coals under stars

twelve policemen dead in the snow

their red jackets
petals of a scarlet flower

pile of bones
GIMME 48 HOURS

Ironhorse farts and shudders,
pisses between the tracks.
Honor, disguised as the conductor,
punches my ticket, kicks me to the floor;
Decay with a crescent moon butcher knife
swipes wildly, the threads in my pants
popping at the groin.

•

listen
listen here's
a good one:

manitou painted this hill and this valley
listen manitou painted the buffalo
before that he was white
so white you could not see him
nor the trees nor the rivers nor the heavens

manitou took up his brush listen
to this he took up his brush
to paint the buffalo and the buffalo grass
the buffalo licked his hand in gratitude
the green grass bowed its head and wept abundant seed

and the buffalo multiplied over the bowl of the earth
manitou painted the cornflower and the rainbow
listen to this listen before when all was white
in the month of drums listen manitou was not afraid
in the month of drums when all was white

manitou took up his brush to paint a world its colors
listen and every pretty girl her lips and her hair
and flung his paint pot to the four gates of the wind
so the pretty girls listen to this
would have flowers for their hair

listen manitou did this and was not afraid

Honor buys me drinks
and black cigars,
introduces me
to his sexy sister.

A hot toe
of flame
slides, snakelike,
my pant's leg.

Last call!
stumbling, stars whirling,
snow cold berth,
her hot nest

—the fingernail
between my breasts
flashes like a knife
and cuts my flesh.

I grab her face,
rip free
the rubber mask:
Decay. Medusa.

•

The flowers of the Indian paintbrush,
unabashed and profuse, bloom
their sharp red cuts into the face
of the pre-Cambrian rock.

Ironhorse portages thru
the gargoyle, stunted trees
whose precious seed will not sprout
twenty miles about Sudbury, Ontario.

O, Champlain, *premier canadien*,
here is your Gougou monster
the savages told you
kept his supper toasty,

six men in a side pocket.
Gougou vomits trade and commerce,
whisky, chips and gravy
the color of nickel slag

and a breed of men and women,
looking for work or a guttered butt,
tough as a stunted pine
with a tenacious grip

on Gougou's dinner napkin.
Chicken wing, chicken wing,
I ain't no chicken wing
bubbling in no chicken pot.

•

(Lone signalman walks, kicks
swims thru two feet
          of snow drifting,

—searching for a broken bond
in the flash signals.
Service call: 3:30 a.m.

Walking a mile in it,
lost and confused—
          the Horn!

Bright light, barely
got off tracks: his heart
running with the hare.)

Whirling flashers. Red and Green.
A white light like a gyroscope
plows the black night.
The dark cows lope,
ass-swish, to dim barns, towns burn
like fuses thru the phosphorescent suburbs.
Everywhere the electric city.

I ride a bead of time (illusion),
am the eyeball suspended
in the steel eye winking, melting
into one past and future—
tree at my window (subject, object)
swift as shutter.
The veins of my palm
or river valley: Northwest Passage
sniffing out the Silk Road
thru impenetrable snow, tho reliable sources
say the Sioux speak with a Chinese accent,
a fortnight's march to the sea.
Monopolists pick their teeth.
Black wolves chew thru a forepaw.

•

. . . and Ironhorse speeds
into the vulva
of Montreal.

A blackjack hits my head.
Angels flutter
with flute and harp.

. . . awake peering thru the threads
of a burlap bag tied at the neck
with half-inch chain
a subway burrows
beneath the river
pulls to a stop
bump bump bump
I'm dragged out and up
Honor breathing heavily
thru fake black moustache
and Jimmy Durante pink plastic nose

Decay limps behind cursing thru a menthol cigarette
leads Treason by a leash
of carved bone. . .

We wind upward (the jeweled dipper
ever nearing) the cracked,
blistered, concrete thread
of a screw.

bump bump bump
I piss my pants.
Honor clips a clothespin
over his plastic nose.
Decay kicks at the wet burlap,
smack of leather boot catches me
in the mouth knocking
loose a front tooth.

Honor dumps me out.
We stand atop a tower
looking out across the St. Lawrence River,
les iles de Montréal, *the firefly huts*
of a metropolis
obscured by a whirlpool galaxy
set like an eye
in the tornado of North America.

Honor and Decay and Treason
surround me
in a triangle.

"Jump," cries Honor.

"Who are you? F.B.I.?"

Decay grins insanely,
her left thumb up her asshole.
Treason sucks a split skull.

"Jump!"

"C.I.A.?"

"Deeper than that, punk. Rush him!"
The three of them charge; I drop,
roll free.
            Their heads crack
like billiard balls.

I leap up and grab Honor;
I wring his neck like a chicken,
ring it like a bell.

FOUR POEMS

Boston Pops

How pretty the little boats
with their little white sails
on the river Charles
but so confused, distracted

in the cross currents:
Big Thunder shakes the land.
It's the 4th of July,
Arthur Fiedler and the *Boston Pops*

brass the *Star-spangled* on the esplanade.
A city shines upon a hill
beneath the fireworks glitter
of ten thousand shattered centers.

I walk up and down in Beantown
searching for the center
of her hive, but don't reckon
I'll find sweet fuck

tho Old Boston Court House
is kept intact,
a squat black broody hen,
"to instruct the patriot in youth."

A story goes that Justice indeed
one day passed thru town,
a peasant girl in burlap
driving her father's swine to market.

She carried in one hand
a long willow switch,
in the other a stone crock
of her mother's honey.

She looked in at a silversmith's
who set aside a precious bowl
to cry thru town
with lantern upon horseback.

But that's all bullshit.
Paul Revere never rode that horse.
Barbecue chicken
spins in a yellow shop window

where I see Sacco... *no! no!*
Yes, I see a fish-peddlar, a cobbler
on the rotisserie! I see—
GET A MOVE ON, BUM,

I'LL CALL DA COPS! YR ASS!
—a fat man with grease
on his fingers,
blood on his apron

wiping my fingerprints
from the glass.
Overhead, the M.B.T.A.
makes a beeline to Cambridge—

Snowbound under glass
upon honey-color paper;
a skinny kid in the streets,
SPARE CHANGE FOR MY VERSES!

Fresh black mimeo ink
runs on his lips and tongue.
Laffter spills out of bars
bright as honey.

VALLEY OF THE DOLLS/
PLANET OF THE APES
Bank vaults
harpoon the sky.

A fortune teller *taps*
tapping on a thin pane
beneath the scaffolding
of New Boston.

Two bums chatter
on a park bench
in the dark Commons.
Listen to the trees cough,

flowers are so dirty.
I lie in the wet grass
as the street people drift,
—"hash", "acid", "speed"—

my face a scrap of bread
tossed toward the pigeons.
An alligator
dances with a pink nude

big as a balloon.
They waltz on roller skates.
Tethered to a broken plow,
the skeleton of a mule

crumbles in the dust of a sidewalk.
Death helps Adam plow,
each taking a turn
round the mirror pond.

An egg-man crawls on all fours,
a dead tree
grows out of his back.
A crowd of men

sometime one in chant
WE ARE ONE / WE ARE
ONE / SON OF A GUN
sometime confused

by other's elbows and wives,
coats and emotions,
climb a ladder
hoisting a huge knife

into the hollow rear-end
of the egg-man.
Wearing the half-moon
for a riot helmet,

a mounted policeman
rises upon a black horse
against the black night,
MOVE ON! MOVE ON!

I roll in the grass
holding my head between my hands,
NO SLEEPING IN THE PARK!
The seed of the dandelion blows

til foothold taken
in the crack of a civilization,
the saxifrage will cut
thru rock; love isn't smashed

or sacrificed complete, rubs
against time rather than with it—
the desert may not flower
for twenty or for thirty—

a thousand years it may not,
but the first dive down from Greyhound
there's an ancient stripper
shaking her can

to the beat of a wild band,
LOOKIT THAT BITCH DO THE DIRTY DOG!
A mountain of a woman,
she belts out cunt'ry blues:

A HANK OF HAIR, A PIECE OF BONE
I blow my last buck
on a cold beer
and bum a smoke.

Wild Flower New Hampshire

If I could know you,
if I could know you intimately

as a woman,
as I would know a woman

being a man
knowing intimately a woman,

holding love to name a flower—
it would be nothing.

•

You gave me a handful of wild flowers
bloodroot and had me smash them
in my fingers, bloodroot stain
on my fingers to know their name
crushing out their brief life
so I bruise your thigh
black and ugly purple

with my fist hitting you hard
to know your name like a mantra
against the dark
taken root in my brain

mantra mantra
bruising your thigh
to know aloud your name
pain, pain you cry aloud
and are afraid, your name
bloodroot smeared
in the palm of my hand
squeezing out a name
mantra of pain

•

Walking the jeep road
up the mountain, we rested on rock
cropping out in a grove of birch
to catch our breath.

I thot I'd like to have you there,
fuck you against the rocks:
you could see the afternoon sun
drop into a grove of birch

and jerk my come into your snatch.
I'd turn you over to fuck again,
then eat out your cunt
while you sucked on my nuts.

But, no—you saw flowers,
pale, white, sickly things
you held out to me and called
their name, "*Indian pipes*,

look here!" and they withered
at the touch, turned black
when picked, put a kind of pitch
to my hand.

•

A great glass pot of cold water
fresh tiger lilies
next a writing pad on the kitchen table.

You sleep, I smoke opium
but I am not casual,
nor nonchalant, nor unrattled,
nor gentle...

I cannot write your name.
Cool midnight
thru the open window,
and it is nothing.

•

We would never see each other again,
but we did not cry out

because we had roses
which we pulled apart petal by petal

until our loving bed was a smear
of hundreds of rose petals.

Then, we had no roses
but the fragments of roses

and did cry out
in each other's arms, "lost love..."

•

The next day we left
after picking blueberries at dawn.

We pulled off the interstate
at Derry.

You asked the gas jockey of Frost.
"Won't for two weeks, ma'am..."
I loved you then for offering him
blueberries in the confusion.

It only confused the man
who wanted money for gas.
We laffed all the way back
to cloverleaf and interstate.

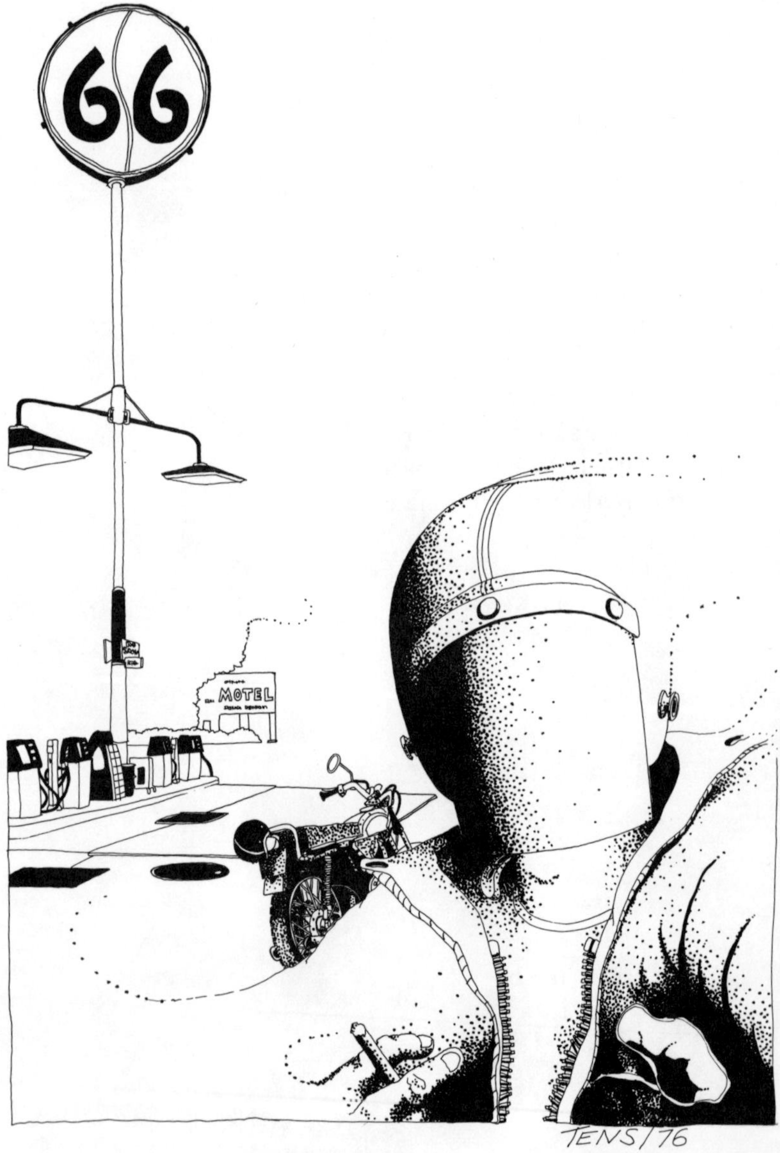
66
MOTEL
TENS/76

Kokomo Crossroads

"The cross is bending,
midnight passing,
and joy cometh with the dawn."
—Eugene Debs

10 miles from Kokomo,
30 miles from home
and I come to the crossroads
of the nation
where fear of the wolf,
the savage tomahawk
left a permanent stutter
to the native tongue:
Who— who— who—
Who's there?
shot into the dark,
fists pound the board.
Wee-yoo, let me in!
I'm bawling like a little baby;
an hour ago,

I said good-by to a lady
—let her out the door—
she's too tired
to be running anymore.
10 miles from Kokomo,
30 miles from home,
3,000 miles to go.

Tell me, who's there?

X-ROADS OF AMERIKA,
foreclosing the ancestral homesteads
was a cakewalk.
Let the people eat shit.
Turn away the sons and the daughters
for the whores of the marketplace
and the cocksuckers of industry.
Smash anyone with a heart,
a split sack of barley mash
along the railway tracks.
The lamplights shining.
Foul the air,
brand the land *K K K*—
your Grand Dragon
chews the breasts of your innocence
til the nipples
stand out bloody red roses
and he spits out
slips of pork fat.

Tell me, who's there?

Robert Johnson mixes Alabama calico,
Chi-town steel
to know your crack-up
crossroads song:
beats out mean Indy blues
on a cardboard, tenement guitar.
Dreiser faints
in a bowl of tepid soup,
the cuffs of his broken suit
swelling like dumplings
of tainted grain and hopelessness.
James Dean plays pick-up basketball,
skips stones cross Cicero Creek,
turns to Jesus
—holy rolling—
trombone slides
jubilant Baptist communion:
dead at 25,
burned out on the road
to nowhere.

Tell me, who's there?

Young boys
with corn-stubble chins,
girls swaying
as yellow wheat sways

before a storm—
all go running.
John Dillinger.
Bill Vukovich.
Lady with that red
dress on. . .
Machine-guns out of alleyways,
Offenhausers climb
the retaining walls of death.
All go running
from the rainbow-parts factories,
the prisons of dawn,
the red and green armies—
70 miles an hour
smack into dogwoods blooming.
All go running.

Tell me, who's there?

Little Orphan Annie
comes to my house to play:
sweeps away cobwebs,
loafs all thru the day—
WIN WITH WILLKIE,
my Dad was still a boy.
50,000 he claims
about the bandstand
and along the river
to cheer big guns,

pop corn and politics.
Labor Day picnics,
turning, whirling leaf—
and on this crossroads morning
I slide thru Kokomo, Indiana,
searching the rush hour tornado
for the Old Man's car.
He's a ham radio nut,
knows Morse Code:
I take a chance, flash
my hi-beam *di di di di di di*
and *di di di di di di*
flashes back.

I'm on the run, who's there?
Hoosier, quick!
Where's Deb's Square?

Flight

White stones, black stones,
the shaved heads of the draftees
carved from hammered bone
whirl and glitter

thru the blazing
San Francisco noon:
the sun a bright bloodclot
at the jugular vein of day.

Lady dark as sorrow,
heavy with child
and twelve-string guitar—
Let my people go.

"Don't mourn for me,"
sang Joe Hill, "Organize!",
moments before the firing squad
turned him beneath sunflowers.

Dance round the pyre
of the flaming dreams,

gargoyles grimace and fade.
White ash on the wind,

I've a ticket to ride
first class 707;
at sundown today
you can kiss my ass good-by.

•

Amerika,

I'm flying the coop,
the whole rat's nest
of your promised land
already promised to the few

promising the gun
and the lie,
the sack of shit,
the sack of lime.

I'm turning myself
inside out
with bitter laffter,
quiet tears.

•

I've got you by the balls

Shall I dance with joy?
Release assassins?
Smash cannibal machinery?
Piss upon the flag?
Whoopee, whoopee doo?

I'm lighting your long black cigar

Here's to the Forefathers
who gave us the Will
and the Freedom
to trade and to trust
in the spilling of blood.

swilling your hard shot of corn whisky

Here's to the Boones
who slit the throat
of the Alleghenies;
here's to the Rockefellers
slitting our purses.

eating your prime sirloin

Here's to the Pinkertons
threshing the mob—
equal justice to none;

here's to the Clays
carving rhetoric into skulls.

fucking your classiest whore

Here's to the Grants
who'd rather kill than maim,
maim than drink;
who'd rather drink
and trash the shitcaboodle.

I'm driving your sleek black limousine

Here's to the bloodsuckers,
the money hoarders,
the war mongers,
the parasites, the goons:
The Last American Poem. So long.

Born in Ohio in 1946, Charles Tidler is a naturalized Canadian citizen. He lives in Vancouver with Carol Tidler and their young son, Sam. *Flight* is his fourth book of poetry.

— PULP BOOKS —

Leo Burdak, *Gearfoot Wrecks*, 75¢
Brian Carson, *A Dream of Naked Women*, 2nd Edition, $1.50
Dan Dougherty, *The National Hen*, $1.50
Roger Dunsmore, *On the Road to Sleeping Child Hotsprings*, rare.
Ken Eisler, *Inchman*, $1.50
D.M. Fraser, *Class Warfare*, 2nd Edition, $8.95 & $3.50
Jon Furberg, *Jonas*, rare.
Bill Hall, *Incest; Breaking the Tabu*, $1.00
Rosemary Hollingshead, *3 Men on Horses*, $1.95
Augustin Hamon, *The Psychology of the Anarchist*, 25¢
Alfred Jarry, *Ubu Rex*, Translation by David Copelin, rare.
David Jenneson, *The Helping Hands of Chirstmas*, $1.00
Chris Johnson, *Duet for a Schizophrenic*, 75¢
John Kula, *The Epic of Gilgamesh as Commissioned by Morgan*, 75¢
Mark Madoff, *Paper Nautilus*, 75¢
The Patient Renfield, $1.00
Carlos Marighella, *Minimanual of the Urban Guerilla*, 75¢
Tom Osborne, *Please Wait for Attendant to Open Gate*, rare.
The Reamers Car Club Blues Band Story, rare.
Brian Shein, *Theatrical Exhibitions*, $8.00 & $3.50
Richard Simmins, *Sweet Marie* $1.50.
Roy Stone, *The Destruction of Vancouver*, 75¢
Johnny Tens, *Tenth Avenue Bike Race*, $8.00 & $4.00
Charles Tidler, *Whetstone Almanac*, 75¢
Flight: The Last American Poem, $2.50
Rick Torch, *Lust Lodge*, 75¢
Vicar Vicars, *Crimes, or I'm Sorry Sir, but We Do Not Sell Handguns to Junkies*, rare
Tom Walmsley, *Rabies*, $2.00
The Workingman, $1.00
Lexington Hero, $2.50
Mark Young, *Brother Ignatius of Mary*, 50¢

$3 PULP, An Anthology of Subversive Writings, $3.00